防拐骗走失

王大伟 著
雨青工作室 绘
[加]Jennifer May等 译

·北京·

作者简介

王大伟　中国人民公安大学教授，教育学博士，博士生导师，技术一级警监。享受国务院特殊政府津贴。

我国首批派往西方学习警察科学的留学生，现代警察科学的引进者之一。出版过《英美警察科学》等65部著作。从教近40年，主讲的《公安学基础理论课》，被评为全国公安系统首批国家级精品视频课程。

央视《今日说法》等栏目嘉宾，参与制作电视节目20余年。创造平安童谣、平安童话、平安童操。长期从事公益安全教育工作，多次深入边远基层为留守儿童和打工子弟免费进行安全教育。被老百姓称为：“说歌谣的警察”。

前言

“送你一支小灯笼，平安童谣记心中，记得有人祝福你，默默送你去远行。”孩子，下雨时有人愿为你撑起雨伞，危险时有人愿把你护在怀中。可是你们总要长大，要独自走夜路，送你一支小小的灯笼——一本平安童谣，让它伴你远行，护你平安。

孩子，你来到尘世，父母、师长、亲友不能永远相伴，但你要记得，他们会永远站在你身后，注视着你长大。不仅是他们，社会上还有很多善良的人，当你需要帮助的时候，请大声开口向他们求助。最后，祝愿每一个小朋友都能平安、健康地长大！每一位爸爸妈妈都能快乐、无忧地养育孩子！

童谣中的小案例，全部取自生活，触目惊心。成人参考，幼儿慎听，保护孩子身心健康。祝福！

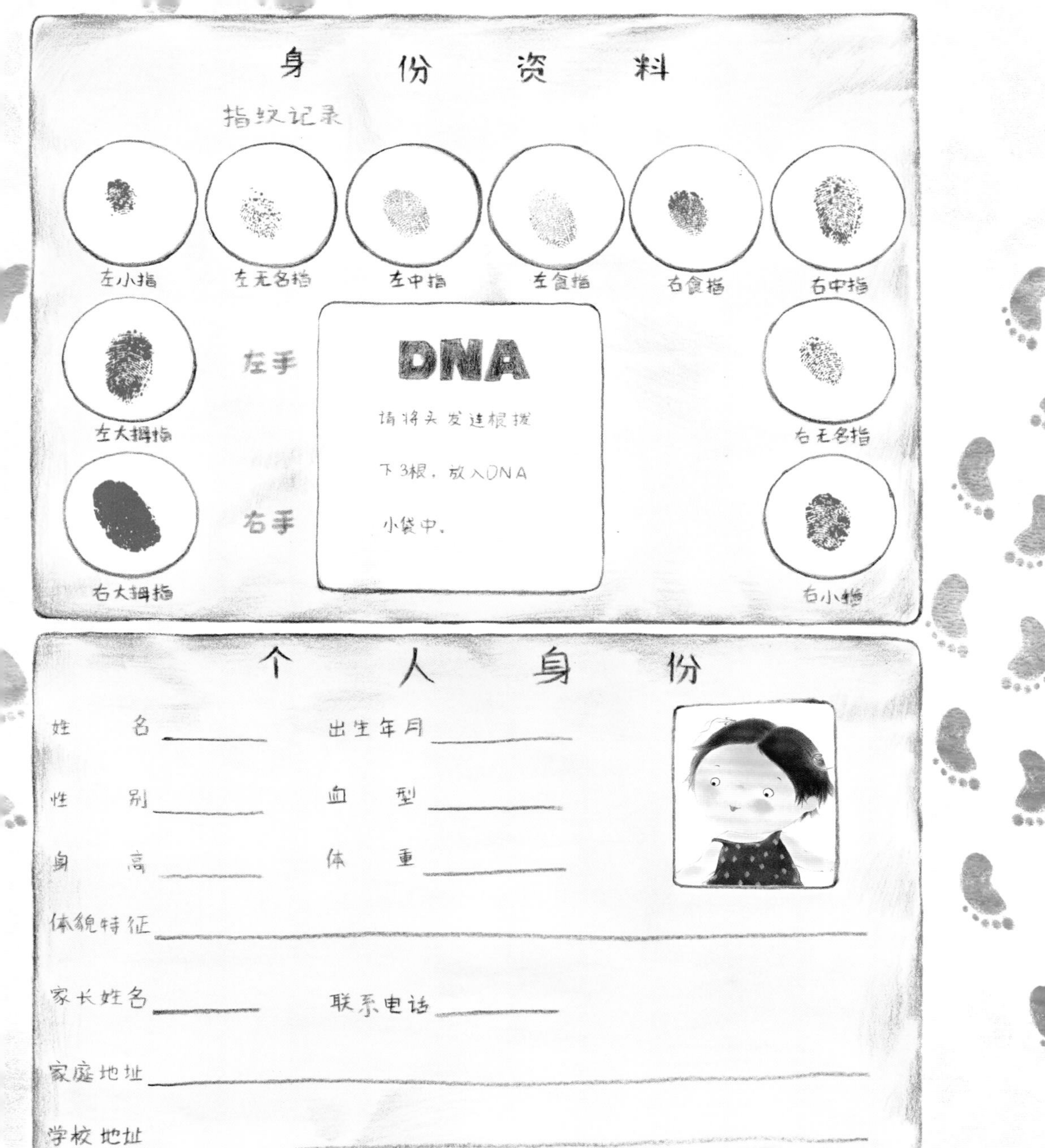
身份资料
指纹记录
左小指
左无名指
左中指
左食指
右食指
右中指
左大拇指
左手
DNA
请将头发连根拔
下3根，放入DNA
小袋中。
右无名指
右手
右大拇指
右小指
个人身份
姓名
出生年月
性别
血型
身高
体重
体貌特征
家长姓名
联系电话
家庭地址
学校地址
联系电话

身份识别卡

小猴子，骑红马，

母子做个身份卡。

十指指纹印卡上，

每个小猴不一样。

小案例
王大伟提示

Little Monkey Card

Red ink on my little fingers.
Make an ID card for me.
Ten fingerprints on one card;
It’s as easy as 1, 2, 3.

Make a little ID card for the children in case of emergencies.

学 会 留 下 小 标 记

遇 到 坏 人 掉 进 坑，

手 里 有 啥 往 外 扔。

书 包 手 套 或 眼 镜，

爸 爸 妈 妈 认 得 清。

Little Signs

When you fall into a hole or pit,
Parents look for signs that fit.
Schoolbag, gloves, and glasses;
Throw these out, even your mitt.

The principle of self-rescue:
Learn to leave personal marks.

剃头拐孩子

超 市 丢 了 小 宝 宝，

长 发 剃 成 小 秃 瓢。

脸 蛋 脏， 衣 服 小，

对 面 遇 到 找 不 着。

Kidnapping

I won’t get lost in the grocery store.
Kidnappers change my looks, therefore,
Not cared for and shaving my head;
I can’t be recognized anymore.

Traffickers will put make up on the children after they kidnap the kids so that their families won’t recognize them.

大灰狼骗小孩的招术

叔叔阿姨在问路，

小狗丢了请帮助。

送你糖果和玩具，

带你去看小动物。

Persuading Children

Four ways adults try to take me away.
Ask directions to know the way.
Candy, using animals and toys;
I won’t go or trust what they say.

***If an adult asked you to help to find a dog.
Just don't go.***

胖小子坐门墩儿

胖小子坐门墩儿，

要带我走可没门。

问我名字还没起，

给我吃的不稀奇。

小案例
王大伟提示

Don’t Talk to Strangers

I am outside, and a game-changer,
I can’t talk to a stranger;
No food or give my name,
Because this means danger.

Say no to strangers.

十大后悔莫及

水火无情切记牢，

眼耳鼻口保护好。

交通安全防狗咬，

出门牵手远人潮。

高楼安装防护栏，

吃的喝的上心瞧。

小案例
王大伟提示

Ten Regrets of Children

Drowning and traffic accident safety to stay alive.
Lost and playing with fire safety tips we strive.
Escalators and putting things in your eye, nose, and ear.
Falling from high places and dog bites, have fear.
Electric shock, makes you shake.
Food poisoning gives you a stomachache

Please remember and try to prevent the accidents.
Otherwise parents will regret it.

春节外出安全

正月十五月儿圆，

赛灯舞狮跑旱船。

拥挤焰火少向前，

不让儿孙离身边。

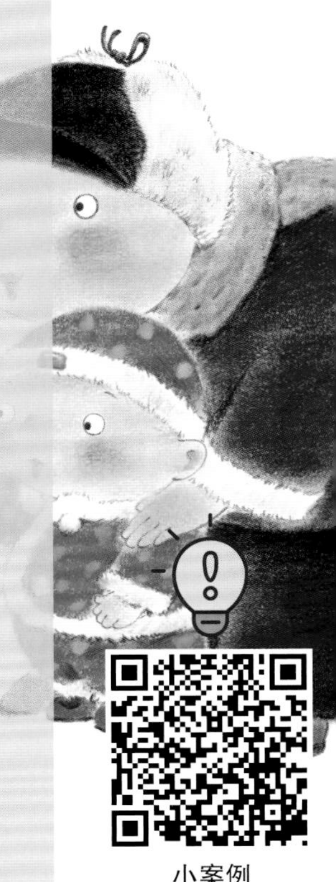

小案例
王大伟提示

Lantern Festival

Holiday events, it’s a must,
It’s crowded and rushed,
Avoid the big crowds;
As I could get crushed.

During the Lantern Festival, try to keep the children away from the chaotic crowds so that we can minimize the probability of stampede accidents and lost events.

婴儿车第一定律

上电梯，下电梯，

孩子拉紧手不离。

先抱孩子后搬车，

丢了孩子悔不及！

Baby Carriage

The elevator, I'm not free.
My mom can hold me.
Then move the carriage off;
I am safe initially.

To keep the baby safe, parents should hold the baby first then move the baby carriage when getting off the elevator.

小熊

小熊小熊好宝宝，

背心裤衩都穿好。

里面不许别人摸，

男孩女孩都知道。

Little One

I am a baby bear, baby bear,
Covered by vest and underwear;
I am only one-year-old;
No one can touch me there.

No one is allowed to touch a girl's body, especially those parts covered by vest and underpants.

犯 罪 月 历 表

较 为 平 安 三 月 三，

四 月 五 月 往 上 窜。

夏 季 多 发 强 奸 案，

冬 季 侵 财 到 峰 巅。

Robbers

Spring and Summer, we adore;
But lock your windows and door,
As the robbers go out;
Safety means more.

The type and frequency of crime will change with seasons.

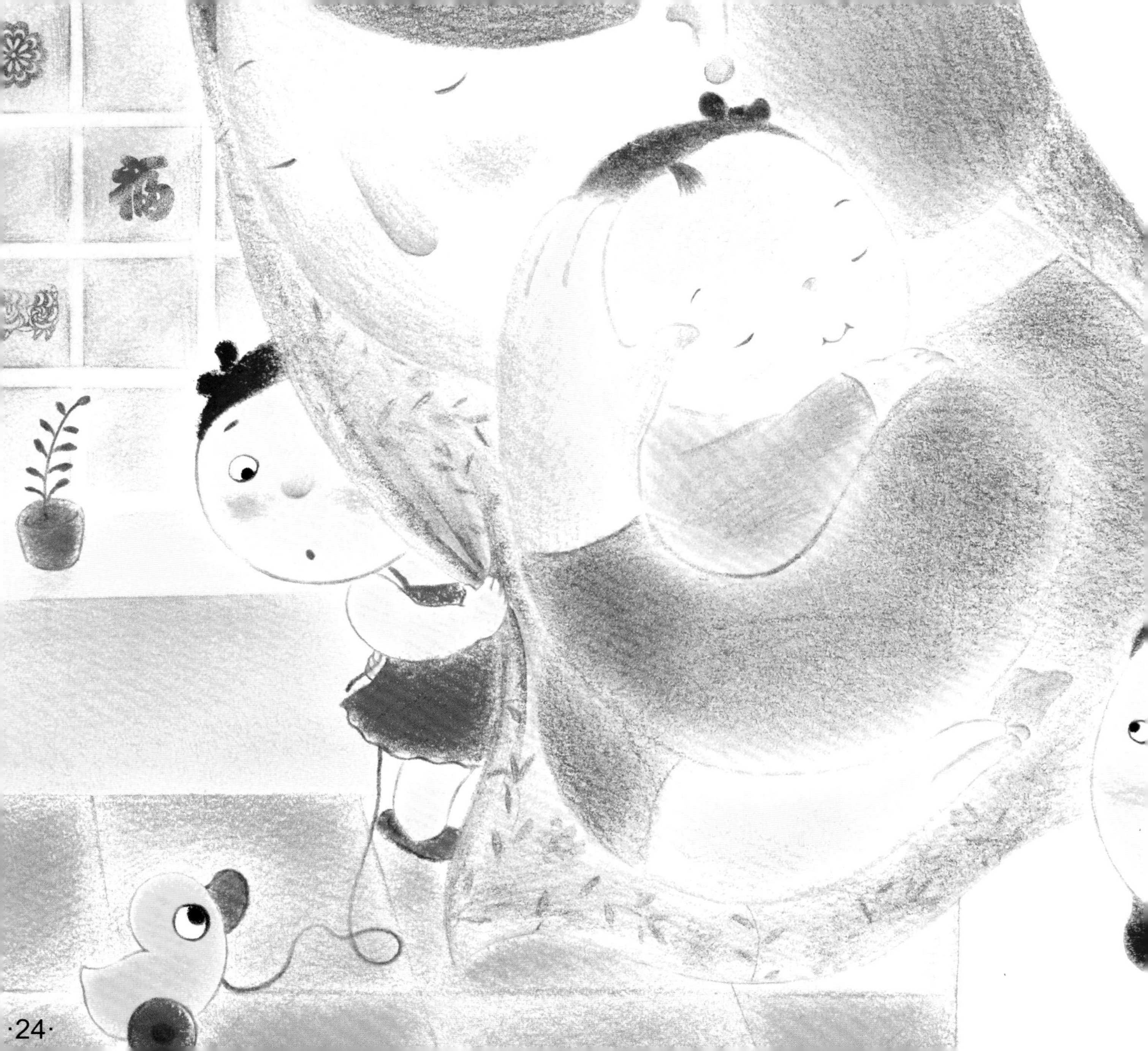
福

女孩安全要诀

背心裤衩不许摸，

慎坐别人顺风车。

小小秘密告妈妈，

问我名字不能说。

小案例
王大伟提示

Tips for Girls

I am the only one to touch my clothes, you see.
I am a girl and I learn at the age of three.
Take no rides from strangers, or give my name.
I will always tell mom and she will protect me.

For girls, sex safety education should begin at the age of three.

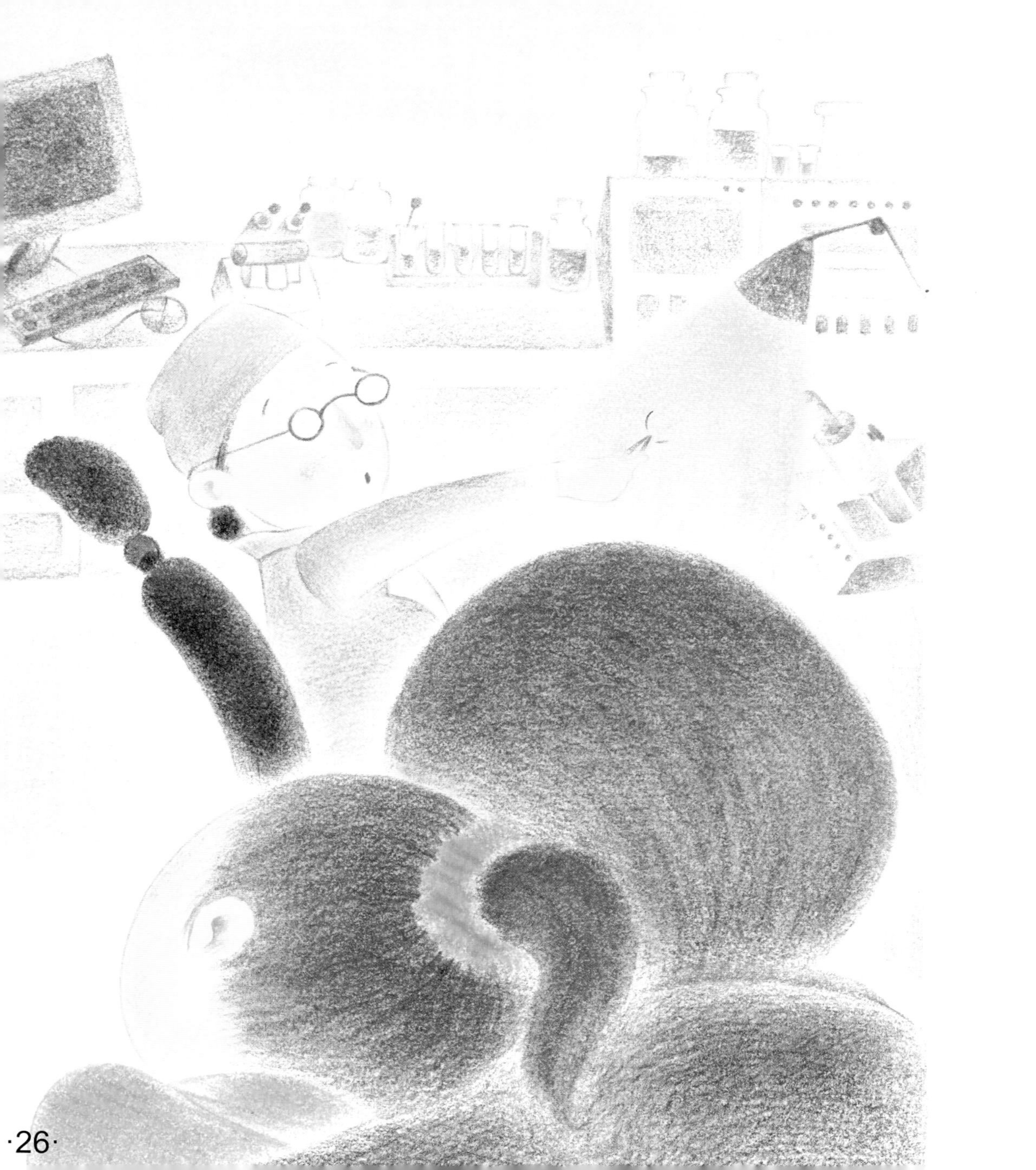

被性侵害后的救助

保守秘密找医生，

毛发体液要取证。

受到侵害告亲人，

平复心理再报警。

小案例
王大伟提示

Sexual Abuse

If I get abused and with no delay,
I will tell my mom right away.
Keep hair, fluid, see a doctor;
And report to the police, that day.

Tell the children the methods and principles of victim assistance.

防走失拐骗歌

你拍一，我拍一，

出门穿件大红衣。

你拍二，我拍二，

不吃生人羊肉串。

你拍三，我拍三，

妈妈电话记心间。

你拍四，我拍四，

问我叫啥没名字。

你拍五，我拍五，

拔腿就跑小老虎。

你拍六，我拍六，

找到警察会求救。

你 拍 七， 我 拍 七，
人 多 拥 挤 咱 不 去。
你 拍 八， 我 拍 八，
自 已 做 个 身 份 卡。
你 拍 九， 我 拍 九，
生 人 叫 我 我 不 走。
你 拍 十， 我 拍 十，
回 家 的 路 我 认 识。

Anti-kidnapping song

You take one, I count one, wearing a red coat, and well done.
You take two, I count two, don't eat strangers barbecue.
You take three, I count three, mom's phone number I can say.
You take four, I count four, my name is the secret to you all.
You take five, I count five, run away like a kangaroo.
You take six, I count six, ask for help, and call the police.
You take seven, I count seven, stay away from the crowded place.
You take eight, I count eight, make an ID card, do not wait.
You take nine, I count nine, say no to the stranger and I will be fine.
You take ten, I count ten, the way to home, I have learned.

包裹一：王大伟安全提示与案例

包裹二：防走失身份识别卡下载

包裹三：能动脑动口的阅读宝盒

包裹四：平安童谣纯正英文领读

扫一扫 即刻查收！

图书在版编目（C I P）数据

王大伟儿童安全童谣. 防拐骗走失 : 汉英对照 / 王大伟著. -- 北京 : 中国水利水电出版社, 2021.9
ISBN 978-7-5170-9617-7

Ⅰ. ①王… Ⅱ. ①王… Ⅲ. ①安全教育－儿童读物－汉、英 Ⅳ. ①X956-49

中国版本图书馆CIP数据核字(2021)第092323号

责任编辑 李格（1749558189@qq.com 010-68545865）

书　　名 王大伟儿童安全童谣：防拐骗走失
WANG DAWEI ERTONG ANQUAN TONGYAO：FANG GUAIPIAN ZOUSHI

作　　者 王大伟 著

绘　　图 雨青工作室

英文翻译 [加]Jennifer May　王大伟　陈诗楠　刘原

配音朗读 王许曈　李晟元　郑方允　崔璎峤　侯清芸　吴郁暖　钟璇　郑淑予

出版发行 中国水利水电出版社
（北京市海淀区玉渊潭南路1号D座　100038）
网址：www.waterpub.com.cn
E-mail：sales@mwr.gov.cn
电话：（010）68367658（营销中心）

经　　售 北京科水图书销售中心（零售）
电话：（010）88383994、63202643、68545874
全国各地新华书店和相关出版物销售网点

排　　版 韩雪

印　　刷 天津久佳雅创印刷有限公司

规　　格 210mm×190mm　24开本　5印张（总）　120千字（总）

版　　次 2021年9月第1版　2021年9月第1次印刷

总 定 价 68.00元（全4册）

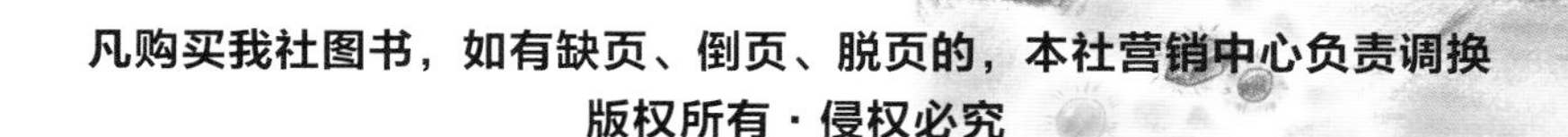